The Moat

The Moat

Human-Centric Customer Experience in the Age of AI

Jennifer Courchaine

Courchaine Media

Edited by Jake Arky

Published by Courchaine Media

ISBN 979-8-9953604-0-7

DEDICATION

For Charles, my favorite human. I couldn't do this without you.

ACKNOWLEDGMENTS

Relationships built this book.

Special thanks to the friends, colleagues, and strangers who replied to my cold email when I reached out. You proved that human connection still scales.

Thank you to my customers, past, present and future. It's an honor to know you, and a privilege to serve you.

Contents

AUTHOR'S NOTE

I started writing this after sitting in an investor meeting where I watched a CEO get pummeled with no less than 20 variations of "What's your moat?" as he conducted his 30 minute pitch.

The pause after the question got worse every time. No answer was good enough, defensible enough, or unique enough to quiet investor concerns.

Every founder I talk to is going through some variation of this soul searching right now. When I look to Customer Experience leaders, they are too. Everyone seems to be going through the same existential crisis. Just when we've finally established Customer Success as a legitimate department, it feels like it's been immediately diminished, automated, or moved under the office of the CRO because revenue is now the only thing that matters.

I had this crisis of faith too. I've been leading customer teams for two decades, and felt like I had a strong understanding of what good looks like for Customer Experience teams and executives, but suddenly I was asking a lot of questions. What does my craft look like in this new world? Is my work even still valid? After a lot of soul searching,

I've realized that fundamentally, the work of caring for our customers still looks the same, at least at the companies I want to help build.

I believe we all need to be more selective in which companies we create now, because there will be two styles moving forward. There will be those who want to use new technology to make things cheaper, and there will be those who want to use new technology to make things better.

These themes converged all at once for me, and I feel real urgency on now being the time to shape our new future. This book could have been a series of blog posts, an essay, or LinkedIn fodder, but the topic feels both important and timely, and I wanted to memorialize it more meaningfully. It's not just another business trend, it's a real decision point on who we want to be.

As I framed out the final content, I pulled out all my business books, and I realized...I've never actually read any of them cover to cover. Every workshop, class, and management program I've ever gone through has focused on the big "So What" in a book, and the rest is just *there*. That's why I stopped writing this at 9 chapters. I want this to be readable start to finish, even though I know how busy you are.

You've got enough on your plate, I don't want to add a 300 page textbook to digest, I just want to help you form an opinion so that you can confidently answer next time the board asks about your moat.

Start this on your flight from Seattle to SFO. You can be done by the time you land.

Thanks for being here.

Jennifer

Part I

The New Competitive Advantage

Chapter 1: The Vanishing Moat

AI Killed My Startup

It's Tuesday. Jane wakes up to a situation that is becoming increasingly common, yet still feels like science fiction. She built a startup, and she built it fast, using all of the cool AI enablement and coding tools that were at her fingertips. Best of all, it was working! For months, they had steadily gained hundreds of paying clients. Their product was specialized, efficient, and solved a real pain point.

Then, it was over.

The big Generative Artificial Intelligence players launched features that directly competed with their core offering. Overnight, the deal close rate dropped. The extremely specific niche Jane's company had carved out was rendered redundant because of gen AI native automation. She had to pivot immediately, moving from building a specific tool to building complex AI workflows, but she doesn't know if this will work in the long run yet, or what will happen to her existing customers.

This isn't an isolated incident. It's the new baseline. Startup founders, growth company executives, and customer experience leaders are all looking at this new world, and wondering what this means for who we are. Everyone is on a journey of discovery, searching for where we fit, how we will build, what we must rebuild, and what skills we need to develop within ourselves and within our teams to ensure we are able to guide our businesses forward in this new, ever-changing normal.

For decades, the plan for building a venture-scale business was consistent: The playbook was to find a large Total Addressable Market (TAM) which had an unmet need or problem that you could potentially address. Then, you had a couple of options. You could build a product that did something different or better than anyone else. You could solve a unique issue in the niche, however small. Finally, you could build something that might not be unique, but you could build it first.

Once you selected your path, you would build your product and go to market. Then, protected with intellectual property or network effects, you would grow.

The gap between you and the competition was called your moat. It was the reason you could charge a premium, the reason you could survive a rainy day, and the reason investors wrote checks. Because that moat was traditionally defined by proprietary technology, first-mover advantage, or a complex feature set, it meant that competitors simply could not replicate overnight. If you built it better, they couldn't copy it easily. If you built it first, it would take time for competition to catch up. That was the fundamental assumption that drove valuation and strategy. You secured your position through code.

Today, that moat is drying up.

The Mechanics of Commoditization

Why is this happening now? It isn't just that AI is smart, or that it's fast; it is that the barrier to replicating software has collapsed.

In the past, if you built a unique product, let's say a targeted sales enablement tool for live calls, you might have 18 to 24 months of lead time before competitors caught up. Your differentiator was the time required to build a complex codebase, train engineers, and iterate on a niche problem.

I worked with a seed-stage startup in this exact space. They were the only ones doing it right. But because the underlying technology had become so commoditized, existing sales enablement companies could clone their functionality in a matter of weeks. The market didn't wait. The space got crowded so quickly that the startup had to completely pivot. They had to change their target buyer, their use case, and their product entirely before they had even gotten out of the gate.

The mechanism is simple: AI agents can read code, understand intent, and execute workflows. When a large incumbent decides to integrate a "smart feature" into their platform, they don't need to hire 50 engineers for two years. They need a prompt engineer for a week.

This shows up in the workforce, too. It feels like every week there are new reports of large software companies writing off significant portions of their valuation or laying off massive chunks of staff. Numbers are astonishing, it's common to hear of new layoffs in the thousands of employees, and it's not the way it used to be, where struggling companies did it as a way to save the business. Now, it's often profitable, growing companies who are shrinking their teams because AI agents can now handle the work those humans were hired to do. If your value proposition is simply the "tech" or worse yet "automation," you are competing directly with the infrastructure on which you are building. You cannot beat the house on price or speed, nor will it refill your moat.

In a world of infinite intelligence, human connection is the only scarcity that actually matters.

The False Economy of Efficiency

This pressure to automate has led many startups down a dangerous path. To preserve margins in a world where features are commoditized, executives cut costs. They remove humans from the loop. They replace support agents with chatbots. They reduce Customer Success Managers (CSMs) and instead just provide onboarding checklists.

They call this "efficiency." I call it "erosion."

In a traditional software business, Customer Support and Customer Success were often viewed as cost centers. As line items to be minimized. The logic was: *If the product is good, they won't need support. If they do need support, keep it cheap.*

But in a world where Product A and Product B are functionally identical, the support experience becomes your differentiator.

When you remove the human element, you don't just cut costs; you remove the only thing that can actually move the needle in a commoditized market. If a customer calls you, they are often calling because something is wrong. Or, they are calling because they want to get more value out of the tool.

This isn't theoretical; you're living this. If you reach out to support, you know immediately if it's a bot. And chances are you get a canned response that solves 10% of the problem and frustrates you 100% of the time. Personally, I've gotten pretty adept at variations of "let me speak to a human."

Conversely, when a human picks up the phone, and is empowered to do more than just read a script, you get empathy, context, and at minimum a path forward. A profound emergent behavior that is ever present when working with customers is the simple power of connection. Just the shared experience of a human talking to another human makes things better. Support might not always solve your problem, but they can always make you feel heard and cared about.

This is the paradox of the current market: Investors are asking for efficiency, but customers are demanding intimacy. You cannot have both at the scale of automation.

The Economics of the New Moat

To understand why this matters, we have to look at the P&L. The economics of software have shifted fundamentally, and the metrics that matter most have changed.

The CAC Trap

Historically, you could grow by spending more on Customer Acquisition Cost (CAC) than you made on the first year of revenue, assuming Lifetime Value (LTV) would catch up. That model assumed you had a unique product that justified the spend, and that once you successfully won a customer's business, you'd keep it for years. It's still true that maintaining a unique value proposition is what makes you special and differentiated. It's what gives you a short sales cycle and years long retention; but with AI commoditizing features, it's increasingly difficult to find a unique value that you can actually hold onto with any success.

The problem is further confounded by the fact that budgets aren't as clear as they used to be. Gone are the days of being able to validate that your buyer had budget, authority, need, and a timeline. With all of the new tech flooding the market, most businesses don't have specific budgets earmarked for a lot of the new technology they are bringing on, so they bring on what they can, when they can. Often companies will run competing technologies side by side for a period of time before choosing their preferred.

So, not only can sales teams no longer rely on feature superiority to close deals, the whole qualification process has blown up. And the more your product is functionally the same as your competitors, the more you have to spend to convince a prospect to switch. The only way to offset rising CAC is to increase LTV.

Churn is the Silent Killer

In a commoditized market, switching costs are low. If you rely solely on your product, a customer will leave at the first sign of a competitor's slight feature edge or lower price. Human relationships, on the other hand, still create "stickiness." People want to buy software from people they like.

It is harder to fire a person who knows your business than it is to fire a piece of software.

If you automate away the humans, you lower your cost to serve, but you likely increase your churn rate. As your churn increases, you destroy the unit economics faster than you save on payroll. You are also likely to see lower LTV and Net Retention Rate (NRR), so you lose the revenue coming and going. This problem compounds at scale.

The Relationship Premium on Exits

When a company exits (via IPO or M&A), buyers look for predictability. Software depreciates. Relationships compound.

I'm reminded of a company I worked with that was solidly middle of the road regarding quality of product in the market. By every technical metric, they were outclassed. Their feature set was behind the market leaders. They didn't have the most cutting-edge AI or integrations, and their uptime wasn't the highest. Yet, when the company went to market, they commanded an acquisition multiple that wouldn't make any sense in a vacuum. Significantly higher than the industry average.

Why? Because they had customer loyalty. Their churn was virtually non-existent because customers felt heard. The acquirer wasn't buying the code; they were buying the relationships. They knew that if they kept the team and the culture intact, they could cross-sell their other products into the loyal base.

That acquisition wasn't a fluke. It was a valuation of this new Moat. In the due diligence process, the acquirer looked at the NRR and saw stability. They looked at references and saw advocacy. They knew that technology could be built in a quarter, but trust takes years. By acquiring, they bought those years of trust in a single transaction.

Valuation Multiples are Shifting

The days of growth at all costs startup models are largely over. Investors are beginning to penalize companies with tech-only product differentiators and reward companies with deep relationship moats.

A $10M ARR company with 5% churn, a Net Promoter Score (NPS) of 70, and human-led customer success efforts will sell for more than a $10M ARR company with 20% churn, a "chat-only" support model, and a generic NPS, even if the latter has better tech.

Your people and your customer relationships as your core differentiation allow you to sustain higher pricing because the customer is buying an outcome, not a tool.

Without Product, what is the Moat?

If product features are no longer the moat, what is?

The Moat is not just good support. It is a strategic asset composed of three distinct, defensible pillars: Trust, Expertise, and Empathy.

Trust: The Liability Shield

AI can make mistakes. In fact, hallucinations are still a core characteristic of LLMs. When a critical business decision goes wrong, customers need someone accountable. They need a human to take the phone call.

Similarly, AI cannot sign a service level agreement (SLA) with a guarantee of outcome. A human team can. Trust is built at every stage, and not only when everything is going well. There are opportunities to build trust even when software fails. Does the human step in immediately to mitigate the damage? Do they communicate proactively before the customer notices the bug?

Your people also add a critical accountability layer. If customers know that in the event of a system outage, a named person answers the phone, this reduces their risk profile.

Expertise: The Consultative Layer

AI is good at answering "How do I use this feature?" It is bad at answering "How do I solve this business problem?"

AI has access to documentation. Humans have access to *context*. They know what worked for Customer A and why, and how it can apply to Customer B.

Expertise will look different in a reactive interaction than it will proactively, but a common thread throughout both is that your customers are relying on you to be the expert.

Generally Support will be picking up more reactive interactions, as the customer reaches out due to a specific need. The Support agent may not have all the details and context instantly, but they need to know how to get it very quickly, and be empowered to use those details and context to solve the problem or meaningfully move toward resolution.

The Customer Success Manager (CSM) is the proactive side of the experience. They need to go beyond "just checking in" with their customers, and instead drive real business value and meaningful conversations. A strong CSM team has the ability to coach. To look at a customer's data and say, "I notice you aren't using Feature X. Based on what you told me about your Q3 goals, you should be because it will solve these problems for you. Here's how I know." That is strategic advisory.

Think of a strong reactive and proactive engagement approach as a pivot not to better software but to better consulting. Asking prospects and customers how they would prefer to utilize the software, and then working hand-in-hand with customers on how to implement the workflows. The value shifts from the code to the advice.

Empathy: The Emotional Connection

Your customers want to like you and be liked by you. This is the hardest thing to quantify at any given moment, but is easiest to measure in sentiment, advocacy, and brand loyalty. Empathy is the ability to recognize a customer's emotional state and adjust the response accordingly, but also to make your customers truly feel cared for.

AI can simulate empathy (e.g., "I understand you are frustrated..."), but it cannot feel it. And customers know the difference. In high-stakes Business to Business (B2B) deals, customers buy from people they like and respect.

In practice, empathy is remembering a detail from a conversation three months ago. It is offering a grace period without being asked. It is going "above and beyond" not because it's in the script, but because it's the right thing to do.

Often, empathy is brought up in the context of the negative, but it is even more powerful when it shows up in the positive. I once led a team who would capture customer promotions, business, and life wins, and we would present them to internal teams and celebrate them. Our CEO was even known to publicly congratulate them. Customers knew we were invested in them not just because of their spend, but because we cared about their personal success.

If a founder realizes their product moat is gone but their relationships aren't, they can now build for the customers lingering unmet needs, leveraging the trust they have already built.

The Pivot to Connection

This book is not a call to stop using AI. That would be naïve. It is a call to stop using AI *as a substitute* for human connection, and start using it *as an enabler* for human connection.

We are entering a period where software will become ubiquitous and indistinguishable. The vanishing moat is the death of the idea that you can win on features alone. But our new Moat comes with the realization that people buy from people.

The following chapters will move from the problem to the solution. That solution may require rethinking how to structure your business so that customer experience is not a cost center, but a retention engine. Your retention engine is more important than ever before, and to support it properly, leaders must look critically at how to fund and empower your teams to be the differentiator, and explore the specific mechanics of building a "Retention Engine" into both your board deck and your daily operations.

If you choose to build faster software, you will lose. If you choose to build deeper relationships, you will own the market.

But before you build, you must understand the landscape. The moat you used to rely on is gone. The question is: what are you building to replace it?

CHAPTER 2: THE COST OF CONVENIENCE

The Illusion of Efficiency

There is a seductive logic that plays out in boardrooms every quarter. It usually starts with margin pressure. A CFO looks at the P&L and sees that Customer Support and Customer Success are the largest variable costs. Then, a consultant or a new VP of Product suggests the solution: automation.

"We can deflect 80% of tickets with a chatbot," they say. "We can replace our tier-one support with AI. We can reduce our headcount by 40%."

The math looks perfect. The EBITDA (Earnings Before Interest, Taxes, Depreciation, and Amortization) improves immediately. The stock price ticks up. The company has achieved "efficiency."

But this is a false economy. It is the financial equivalent of selling off your factory machinery to pay this month's rent. You look great on paper today, but next quarter you have no product. In the context of customer experience, this is the Cost of Convenience.

Companies are trading long-term retention for short-term margin gains. They are prioritizing convenience for the *company* (cheaper operations) over convenience for the *customer* (getting their problem solved).

The Deflection Game

For years, the support industry has operated on a metric called "Deflection Rate." The goal was simple: prevent the ticket from ever being answered by a human.

The logic was that if you could get a customer to read a help article instead of paying a human to answer a question, you saved money.

Chatbots ask, "How can I help you?" (Select from menu) -> "Here is a knowledge base article." (End).

Forums direct you down a single path, "Search our community first." Voicemail delays you, "All agents are busy. Please leave a message."

In a world where software was unique, this was acceptable. If the product was complex, customers expected friction. They bought the tool to save time or increase compliance, not because it was easy to use.

But now, with AI commoditizing the product, the friction is no longer in the usage of the tool, it is in the support of the tool.

I recall a conversation with a VP of Support at a mid-sized SaaS company. They had implemented an aggressive deflection strategy. They bragged about reducing their headcount by 30% while maintaining their ticket volume. They were winning on the internal dashboard.

Six months later, their Net Revenue Retention (NRR) started to slide.

The CEO called a meeting. "Our product works fine. Why are they leaving?"

The answer was in the support data, but not in the metrics they were tracking. They weren't measuring satisfaction–they were measuring resolution volume. Customers weren't leaving because the software broke. They were leaving because when they had a critical issue at 2 AM on a Thursday, the bot told them to read an article. They felt like a number, unheard.

The Hidden Cost of Friction

When you remove humans from the loop, you introduce friction. Friction is the enemy of trust.

When humans are in the loop early and often, you are able to reduce the cognitive load on your customer. A chatbot asking "How can I help?" requires the customer to formulate a problem. A human asking "What is happening with your account?" requires the customer to simply say it's broken, because the human can follow up with clarifying questions.

One of the most clear areas that friction shows up as a negative drag on your metrics is Time to Value (TTV). TTV is typically tracked in relation to initial go-live after product purchase, but you should also use it to track the time until the customer is able to achieve full value. Customers buy software to solve a problem. If it takes them three days to get help because they are cycling through bots, their time to value extends.

If you do not address the friction in your product and service, you also run the risk of escalating brand damage. In the age of social media, a bad support experience is a social media post. A good support experience is an email confirmation at most. Negativity spreads faster than positive feedback.

All of this creates a silent tax on your business. You might have saved $500,000 in support payroll, but see your churn increase by 5%, and losses of $1M in annual recurring revenue (ARR). The math rarely works in favor of the cost-cut.

The Churn Multiplier

Let's look at the numbers.

Scenario A: $10M ARR. 10% Churn. Support Cost: $1M. Net: $9M.

Scenario B: $10M ARR. 10% Churn. Support Cost: $0.5M (via bots). Net: $9.5M.

On the surface, Scenario B looks better.

Scenario A (Year 2): Retention keeps growing. $10M + Growth.

Scenario B (Year 2): The friction causes churn to creep to 12%. The ARR shrinks. $10M becomes $9.5M. $0.5M support savings remains.

The gap has already started to close. By Year 3, the reputation damage means you have to spend more on Sales and CAC to replace the churn. The "savings" evaporate.

The Incumbent Blind Spot: Where the Opportunity Lies

If this is all so devastating, then why are companies doing this? Why do they willingly damage their own long-term value? For some, it's hubris, for others, it's misaligned incentives. It's not that they want to hurt customers; it's that their financial model rewards short-term extraction over long-term value.

I talked to a colleague, Frank, after he participated in an event with several large hospitality-based organizations. There was a panel discussion where they were highlighting all of the new automations and efficiencies they were building into their workflows. He raised his hand and said it sounded like they were setting up a process where they did not care about the customer experience anymore, and was met with silence and some vague hand-waving.

"Jennifer," he said, "I couldn't believe it. They essentially told me they do not care. They don't care about my experience, even if I'd pay a premium for a good one."

It was a stark moment of clarity. That attitude—not caring—is becoming the norm in three specific categories of companies:

1 - **The Giants:** This cohort of companies feel too big to fail. They already have most of the market share. They believe they can't be replaced, so they optimize for short-term margin over long-term love.

2 - **The Niche Players:** Companies that are extremely targeted in the specific problem space they solve. They believe they have your business either way because there is no other option. They treat you as a captive audience.

3 - **The Capitalized:** Venture Capital (VC) or Private Equity (PE) backed companies with executives or boards who prioritize the short-term numbers. They want fast returns at any cost. They view churn as a feature, not a bug, because they can refinance the debt against future growth that doesn't exist.

THIS ATTITUDE IS A GIFT TO YOU.

When large players remove the human element, they lose power. They are actively disarming themselves. They are building a fortress with no front door, let alone a moat.

Once built, organizations have a very difficult time changing, especially at scale. This means that even if these players decide to change their perspective and philosophy, it will take them a very long time to meaningfully change the customer experience for the better.

If you are small: You can care.
If you are nimble: You can fix it immediately.
If you are founder-led: You can treat every customer like they are the only customer.

Just caring can be the gamechanger, provided you've baked it into the core of your business. In a market where the giants are optimizing for cost, optimizing for care is a strategic differentiator. You can win market share not by having a better feature, but by having a better relationship.

This is your Moat in action. It is the ability to say "We see you" when the competition says "Read the FAQ."

The Low-ACV Fallacy: Replacement vs. Augmentation

The biggest objection to this strategy usually comes from finance or operations: "We have low Average Contract Value

(ACV) customers. We can't afford a human for every one of them."

This is the Low-ACV Fallacy. It assumes that scaling Customer Experience (CX) means replacing humans with technology. The Standard Playbook is that high ACV gets a human, and low ACV gets emails, webinars, and maybe a bot. This creates a two-tier experience. The moment you introduce a tiered system, you signal to the low ACV customer: *You are not important enough to talk to a human.* That is the exact friction that drives churn in the segment that has the highest potential to grow into a High ACV account.

Instead, consider your Digital & Scaled Programs as a Force Multiplier.

Do not misunderstand, this is not to say you should assign a dedicated CSM on every $5,000 contract. That is unsustainable. Instead, look at how you can use digital tools to amplify human effort, not replace it.

Replacement will erode your Moat, while augmentation will build it. With standard programs, the customer either interacts with no one, as they consume information only; or they interact with a bot. The bot fails. The customer escalates. The human ignores the escalation. The customer leaves.

When using digital programs and AI to force multiply, you can have your AI monitor account health, give signal alerts, even draft a personalized email. But the human needs to stay in the loop. There is no substitute for human engagement, even if in some cases that's just a quick 30-second review before sending an email.

In an augmented model, the customer always receives a human signature, a human tone, and a human acknowledgment on their specifics. But the human doesn't have to spend 20 minutes writing every email; they only spend the time they need to validate the insight.

Human-Less vs. Human-Powered

By using digital touch and AI to surface the *right* moments for human intervention, you ensure that your humans spend their time where it creates the most value: resolving complex problems and building relationships. The bots handle the scheduling, the data gathering, and drafting for follow-ups. The humans handle discernment and trust. Do this right, and you can scale the impact without linearly scaling your headcount.

The "Human Tax" is an Investment

Loyalty doesn't happen by accident. It is earned. It is acquired with time, attention, and empathy.

When you look at your P&L, do not view the human touchpoint as an expense. View it as a Customer Acquisition Cost (CAC) Offset.

The companies that win in the next decade will be the ones that invest in their people as their Moat. They will accept that their support costs per seat might be higher than the competitor's. They will accept that they might take 5 minutes longer to resolve a ticket because they are ensuring the customer is actually happy.

The Convenience Trap

We live in a world where Amazon delivers in two hours. We are conditioned to expect instant gratification. In B2B software, "instant" has traditionally meant a quick answer. But the new metric of convenience isn't speed, it is outcome.

A customer doesn't want a ticket closed in 5 minutes. They want their business problem solved.

The companies that understand this will be the ones that double down on the human element. They will stop apologizing for taking the time to listen. They will frame their support as a premium service.

The Path Forward

So, how do we move from the Cost of Convenience to the Value of Connection?
It starts by admitting that efficiency is not the only metric that matters. By empowering the people who talk to your customers to make decisions without asking for permission. By understanding that a happy customer is cheaper to keep than a new customer is to find.

But more than that, it starts by recognizing the landscape. The incumbents are retreating behind walls of automation. They are ceding the ground of human connection. Do not retreat with them.

Move forward. Invest in your team. Build the relationship.

This is your competitive advantage.

Chapter 3: Doubling Down on Customer Delight

The Myth of "Good Enough"

In the startup world, there is an obsession with "MVP"—Minimum Viable Product. In practical terms, this often has an emphasis on the "minimum" rather than on being a truly viable product. It is a term that has bled into customer experience. Many founders operate under the assumption that as long as their support is functional, they are safe. Here are some common examples:

Ticket Response Time: Under 24 hours.
Resolution Rate: Over 90%.
Satisfaction Score: Above 4/5.

If your metrics look like that, then you are operating at a baseline of "Good Enough." But in the age of AI, "Good Enough" is actually "Too Low."

When your product is functionally identical to your competitor's, your customer's decision to stay is not based on whether their ticket was resolved in 10 hours or 12 hours.

It is based on how they feel when the ticket was resolved, and whether or not the problem is actually solved. Did they feel like a number on a spreadsheet? Or did they feel like a partner?

This is where striving for Customer Delight comes into play.

Delight is often misunderstood. It is not (just) sending a box of chocolates on Christmas. It is not (just) a "Happy Birthday" email. In the B2B software context, Delight is a human element in concert with proactive value delivery. It is solving a business problem before the customer realizes it exists. It is human intervention that turns a crisis into a victory.

There is something special about infusing humanity throughout your interactions, so look beyond the business side and find ways you can surprise and delight your customers unexpectedly. This might be in the form of welcome packages when a customer signs up for your product, a note on a special holiday, or some thoughtful swag delivered to customers' rooms at a user conference. Build surprise and delight moments into your budget, and take feedback and ideas from teams at all levels in your organization.

CX Starts at the Top

There is a dangerous assumption in many organizations that the Customer Experience is the responsibility of the Customer Success or Support department. This is a failure of leadership.

CX is a team sport. You cannot build your Moat if the CEO does not care about the customers personally. If the leadership team treats support tickets as a cost to be minimized, the frontline agents will treat customers as tickets to be closed. The culture trickles down, not up.

For a true customer-first culture to exist, the executive team must be visibly involved. This can look different depending on the make up of the team, industry, and company, but a common thread is almost always how much your executives actually talk to customers.

When the leadership team is engaged, the message is clear: our customers matter. This alignment creates a unified front. The agents know they have the backing of the company. The customers know that the people in charge are listening.

The Partner Test

To understand why this culture matters more than features, I want to share a specific example from my own history.

Years ago, I was at a company where our leadership team was collaborating to support an account executive who was working on an opportunity with one of the largest companies in the world. We were a small company with a low double-digit ARR. This was set to be our largest customer ever, by an order of magnitude.

But we had a problem. We didn't meet a bunch of their requirements, and we knew a competitor did. On paper, we should have lost.

The customer had a big request. They asked for a specific large data transfer so they could see us at scale. Our competitor took over two weeks to provide a response. No updates while they waited. Just radio silence for 14 days.

We knew this deal could make us. We also knew we couldn't deliver perfectly in time; however, we had something they didn't: a relentless focus on making our customers feel taken care of. It was in the DNA of the business from the time it had started in our founder's living room.

So, the leadership team stepped in to assist. It took us a day and a half of nose to the grindstone work across a couple of people to get the data working well enough to show them. We didn't have the perfect solution, but we had a working one, and we had delivered it with speed and transparency.

We found out after the fact that this was the turning point for us winning the business. The customer told us they knew they were big and complex and that no one was going to be perfect. To our surprise and delight, they weren't looking for perfection. They were looking for a partner they were sure would be there when, not *if*, things went wrong.

That one request didn't win them over on features. It won them over on **trust.** They knew we were a partner who would move mountains for them, whereas the competitor was a vendor who would hide behind a ticket queue.

This is the Moat in its purest form. It is the decision by the leadership team to prioritize the relationship over the process. It worked before AI, too.

We didn't win by accident. We followed a pattern.

The "Above and Beyond" Playbook

So, what does this look like in practice? How do you operationalize "going the extra mile" without burning out your team?

The Three Rules of Delight:

1 - **Proactivity:** Do not wait for the customer to complain. Use your data to find the friction before they do.

2 - **Accountability:** The human is the owner of the outcome, not the process. The process can bend, but the outcome cannot.

3 - **Personalization:** Treat the customer as a human being, not a seat number. Use their name, remember their goals, both personal and professional, and reference their history.

The Empowerment Gap

The biggest barrier to Delight is not budget. It is bureaucracy.

In most organizations, CSMs and support agents are handcuffed. They cannot issue a refund over $50 without approval. They cannot extend a trial without a VP sign-off. They cannot give away a feature upgrade without a manager's email.

This is the Empowerment Gap.

When you force an agent to ask for permission, you force them to prioritize the *company's* internal safety over the *customer's* immediate need. The customer hears: "You are not important enough for me to spend 30 minutes to get approval."

The "Ritz Carlton" Rule for SaaS

The luxury hotel chain Ritz Carlton is famous for a rule: *Every employee has a $2,000 discretion to solve a guest problem.* They don't need to ask a manager. If a guest is hungry and the restaurant is closed, the staff can order pizza. If a guest loses a suitcase, the staff can ship new clothes. The cost is covered. The trust is built.

Startups might not be able to afford a $2,000 discretion, but you can afford something. Refunds and credits are easy to understand, but there are non-discount ways to move the needle, too.

Invites onto Advisory Boards sound counterintuitive, but can truly turn an unhappy customer around by giving them a productive way to share their experience. Professional Services and Consulting are wonderful ways to enable your teams to solve problems and leverage expertise. Premium feature access can also be a huge lever for teams to be able to offer.

Use your Subject Matter Experts (SMEs) and other resources early and often. Let your Support and CSM teams have a threshold of features, and consulting or services projects that are either offered for free or that they can discount.

Personally, I have a standing rule that any customer, regardless of contract value, can be escalated to any executive, at any time. It's almost never used, but it's always there in case of emergency.

The exact options you offer to your team are less important than the fact that you have options to begin with. This empowerment dramatically improves your speed. You'll find problems are solved in minutes, not days. But perhaps even more importantly, it dramatically improves morale. It tells your team, "We trust you."

Employees who are trusted, trust the company back. They stay longer. They perform better. And they take better care of your customers.

Hiring for Humanity

You cannot hire a script reader and expect them to build any sort of defenses for you. You have to hire people who you are willing to let be people, with the discernment that comes with it.

In the interview process, move past the technical questions. You need to find people who derive genuine satisfaction from helping others. Strong fits for Customer Experience team members will have desire for impact and solving customer problems as their primary motivations. Hire right, give them some power, and they will build your Moat with and for you.

Once they are hired, train them on *business outcomes*, not just *product features*. When your team understands the *value* they are delivering, they can advocate for it. They stop being order takers and start becoming advisors.

The truth of the matter is, for the past decade, businesses have underinvested in their customer-facing teams. Support is typically the lowest paid group in an organization, with the least amount of dedicated onboarding, and the fewest options for growth paths. If you want to be a next level company, invest in these people. Pay well, train well, and offer real, meaningful promotion paths that include Customer Success and Implementation, but also Support Engineering, Operations, and Product.

Funding the Moat

Too many CX leaders shy away from the sales or commercial side of the business, but to understand the full picture, you must address the budget. This is where the Cost of Convenience directly fights the Investment in Delight.

Many CFOs view Support and Customer Success as a cost center. They want to minimize the headcount to improve EBITDA. **Stop it.**

You need to reclassify this line item. Your words matter. Always talk about your budget in the context of retention and expansion. It's not just budget spend, it's a quantifiable investment in retention.

Yes, your headcount might go up. Yes, your cost per headcount might go up too. You may need to plan for your EBITDA might take a hit in the short term. In the long term, though, your Gross Revenue Retention (GRR) and Net Revenue Retention (NRR) will jump and more than pay for themselves.

When considering how to spend your dollars across your CX team, the budget allocation should be roughly 60/20/20. 60% toward salaries for empowered, human-centric CSMs and Support. 20% toward Systems and Tooling, including AI and Digital Touch to aid in that team empowerment. 20% to Experiences, including swag, events, direct communication channels, "delight budgets" for solving problems.

Frequently, CX leaders have delegated their power to the technology teams on the software budget, and the marketing team for the experience budget. I suggest having your CX leadership take this power back, and be accountable as true business and revenue focused leaders. Let them own their full budgets to orchestrate a truly powerful experience.

The world is full of trade-offs, so let me be clear: this Moat is so important that if you have to choose between a new feature and a better support team, you should choose the support team. While a great feature brings them in, it is a great team that keeps them there.

Part II

Execution & Leadership

Chapter 4: Your Guiding Philosophy

The DNA Problem

Saying "we care about customers" on a website or in an all staff meeting is not enough. Most companies have a culture that looks good on paper but fails under pressure. When revenue dips, CX gets cut first. When engineering is behind, support gets blamed. When the board asks for margin improvement, support headcount drops.

This happens because the culture is not encoded in the DNA of your business. To shift this, build and evangelize a Customer Experience Philosophy to guide your teams.

A Customer Experience Philosophy is not a North Star. A North Star usually has specific goals around it. This is more of a "who we are" and less of a "where we're going." Think of it as your moral code. If you find a watch on the ground, are you a person who tries to find the owner? Do you leave it where you found it? Or are you a person who takes it? The exact steps you might take will change based on the situation, but you know if you're a person who tries to give a lost watch back or not.

Codifying your guiding philosophy will allow your team to make the right decisions, even when the playbook is silent.

Know Who You Are: Defining Your "Customer Contract"

In product management, roadmapping and prioritization force alignment. You can't have more than one #1 priority without introducing significant thrash, and thrash is expensive at the developer level. Good product people force stack-ranking alignment and show where the cutlines are.

Customer-facing teams, by contrast, tend to organically live in a world of maybes. Businesses are all slightly unique, and because of this they need specific types of customer organizations with specific focus areas. If you have a highly technical product, you likely index heavily on support and professional services or consulting. If you are an acquisition heavy company, you will typically need to have very strong commercial and go-to-market CSMs, and strong implementation and data teams.

There are several right answers, so I am not here to tell you which one you should be, but I guarantee that you cannot be everything.

To get that clarity, you need a way to cut through the noise and ensure that at a high level, everyone knows and agrees on what you are and what you aren't. This allows you to easily prioritize when trade-offs must be made.

Most startups only write down the Commercial Contract, but consider writing down your philosophy as a Customer Contract to make it come to life. No need to overcomplicate it. This should be a simple document, perhaps a sentence, paragraph, or page. Then, when something goes wrong, use it as an opportunity to make your people even better.

Your customer contract can, will, and should vary from other leaders and other businesses. This is to be expected, as it creates visibility into differences and lets you drive clarity, but let's consider a couple of examples and behaviors they might drive.

"No Blame"
"When things go wrong, we do not argue about who is at fault. We ask 'how do we fix it?' and 'how do we prevent it?' "

This mindset might be tested in a scenario where a feature defect causes a data outage. Engineering blames Support for not testing. Support blames Engineering for bad code. The opportunity is then available to reset the behavior. Instead of continuing to shift blame, the meeting starts with "How do we fix the customer impact?" before discussing the root cause.

"We Own It"
"If you have a problem, we do not pass it to someone else. The person who answers the phone owns it until it is solved. If it requires engineering, we escalate it. If it requires a refund, we approve it. We do not say 'that's not my job.' "

A challenge here may arise in a situation where a support agent transfers a billing dispute to Finance without notifying the customer. The customer waits 48 hours with no communication and gets exceedingly frustrated.

Instead of continuing to pass the buck, the manager steps in, and the agent is coached immediately. The process is updated to require a handoff notification. The contract remains intact because the team owns the fix.

The Measurement Framework

To operationalize your customer culture philosophy, you need a measurement framework that goes beyond standard support metrics to ensure your philosophy drives business outcomes.

1. Clarity: Are you making better decisions, and are you making them faster?

Measure: Track **Decision Latency** (time to resolve without manager escalation) and **Escalation Rate.**

If your philosophy is clear, your team shouldn't need to ask a manager for permission on routine issues. Target <10% of routine issues require manager escalation.

2. Consistency: Are customers and colleagues aligned, and do they feel we deliver on our promises?

Measure: Look at **NPS trends, Customer Effort Score (CES),** and **Win/Loss Analysis** specifically tagged to "Support Experience."

Ensure that a customer feels the same level of care whether they are interacting with Sales, Support, or Product. Variance in sentiment scores across departments should be <5%.

3. Communication: Are you showing (and telling) quantifiable impact across the business?

Measure: Tie the philosophy to outcomes like **Expansion Revenue, Lower Churn (GRR/NRR),** or **Increased Product Adoption.**

Communicate specific customer stories where consistent application of the philosophy created trust or advocacy.

Highlight quotes from team members about how the philosophy helps them prioritize their daily work.

Lead from the Front

You cannot expect your team to embody this if you do not embody it first. By that same token, a philosophy is useless if it stays in the CEO's office. It must be distributed throughout the business.

In the early days of a startup, the Founder *is* the Experience. Every email, every call, every decision reinforces the DNA. As you scale, the risk is that the Founder gets distant. You must infuse the ideals throughout the business, but you cannot delegate the ownership of it.

If a senior hire doesn't agree with the company's customer contract, they are a bad hire. When promoting managers, look for those who live the customer contract, not just those who hit the numbers. If someone violates the spirit of the business, they must go, even if they are a revenue star. This is the ultimate signal that the philosophy matters more than the short-term result.

At scale, these commitments are commitments the executive team makes to the customer and to the team, and should be the first thing you ask your employees to internalize. Sales is empowered to sell the relationship, not just software. Engineering can focus on building for outcomes, and Finance can track support as an investment, not just cost.

The first day of work for every employee should not just be about benefits and payroll. It should include your Customer Contract. Include Customer Journeys in your onboarding, too. Make it impossible to join the company without understanding who you are serving, and why it matters.

The Internal Consistency Challenge

The best time to do this is right at company inception. The second best time is now.

If you're starting after the team is formed, the biggest barrier to moving toward a philosophical perspective on how you treat customers is likely to be internal from the customer-facing teams themselves. It is extremely hard to get Customer Success teams to actually think at this level. Even your leaders will have trouble with this at first.

There will be friction. There will be moments when the philosophy clashes with the P&L. Do not automatically say yes to everything. But do not say no without exploring the solution first. This distinction is important to who you are.

The Soul of the Business

This is not "soft." It is strategic. It is the only thing that AI cannot replicate.

Code can be copied. Features can be cloned. But a philosophy, deeply rooted in a team, and committed in customer contract form, is unique. It is the first step to scaling your Moat.

Chapter 5: Building the Retention Engine

The Battle Plan for Defending the Moat

Too many startups hang a flag on their wall that says "Customer Obsession," then build a support queue that treats customers like tickets. This is not hypocrisy; it is a failure of architecture.

The most common failure point in building your new Moat is the organizational chart. Most companies default to a siloed model because it looks clean on a slide. Sales owns the lead until the contract is signed. Support owns the ticket until it is closed. Customer Success owns the renewal until it is signed.

In this model, the customer is passed from hand to hand like a hot potato. They have to re-explain their problem to every new person. Trust is reset with every interaction. This is the experience that you can replace with a bot, since you are not sharing context or building relationships along the way.

The Cost of the Silo

A VP of Product at a Series B startup recently told me that they had lost a major enterprise deal. The product was technically superior to the competitor, yet the prospect chose the competitor. Why? Because the customer onboarding process involved handoffs from three different departments, each with a different tone. Sales promised speed. Support demanded patience. Professional Services demanded data and a tight Statement of Work (SOW) with rigid guidelines for what was included in the scope of the project and what was not. The prospect felt like they were being herded yet not served.

They didn't lose on product. They lost on friction.

The Collaborative Model

There are always alternatives to silos. What makes sense will depend on your industry, product, and stage, but we can simplify the complex structures into two types: Generalist and Specialist teams.

Generalists

This model works extremely well for early-stage startups with low volume and high complexity, or for your top-tier enterprise accounts where intimacy is the only thing that matters. Your Generalist team is a one-stop shop for anything a customer needs. They are the conductor for the customer's music. Yes, they may pull in others within the organization, but they own everything related to the customer and combine technical expertise with soft skills.

In practice, this will be a single, named person for an account, let's call him Brad. Brad is a senior industry professional, who gets involved in large sales opportunities at the late pre-sale stage to make sure a prospective customer is a good fit. Once the deal is signed, Brad starts to implement the software, working side by side with the customer's project teams. Once the customer goes live, Brad continues to be the main contact when they have questions, feature requests, new workflows or functionality they want to roll out. Brad might work with product, engineering, support, and other internal departments, but the customer knows they just call Brad.

Generalists can be excellent for your top 10-20% of customers; you can afford to hire for extreme quality, and it is easier for a customer to build a relationship with a single contact than a whole team. Ironically, Generalists can also be a great fit for your bottom segment of customers, as lower complexity means more overlap in need and communications, so a lot of content can be operationalized and broadly shared with the customer base.

The challenge with generalist teams is that they take real and concerted effort to scale. You need very high ownership and high-skilled people, so cost per hire is high. You risk concentrating knowledge in a single head as well, because the generalist does it all, so they also tend to be a bottleneck on documentation and systems updating. Internal records might not stay up to date without prodding. This is why most companies start with generalists but end up with silos. It is easy to slowly split out the knowledge as you grow, but it doesn't mean that's always the right choice. If you want a truly premium experience for a class of customer, this concierge type generalist approach is extremely powerful.

Specialists

Specialist teams are common at scale, but they are easy to

silo if you don't take extreme care to build them with clear dotted-line connections and ownership between not just Sales, Professional Services, Customer Success, Support, and Product leadership, but also with the individual contributors doing the work across each functional team. B2B SaaS companies will usually have three key teams in their customer-facing organization.

The first team is your project based group. This is your implementation/project management, training, both remote and in person, professional services, and/or consulting. Some of these services may be classic project management to get customers live, while some may be technical, custom, or subject matter expertise focused. These services may be paid or free, but even if there is no cost they will typically be bundled into your overall selling price. TTV will be this team's key metric.

The second team is your reactive group. This is Support, but depending on your organization you may have Basic, Standard, and Premium Support. SLAs will be the standard metrics here, such as Time Per Ticket, Same Day or First Touch Resolution, Ticket Escalation, and Time to Close.

The third team is your proactive customer success management. This is your most commercial focused team, made up of CSMs of various levels, Digital Touch, and likely some Operations support. Key metrics for this team are Gross Retention Rate (GRR) and Net Retention Rate (NRR).

Now that we have these three functional groups, we need them to work together. For each customer, team members should be assigned to work together across each department assigned to a region, customer segment, vertical, or key business niche.

Each person in the specialist model should be accountable to each other beyond interactions in shared customer calls. The group should meet regularly and collaborate as a team, not as separate departments. This structure needs to align incentives, so If the customer takes a year to launch, or churns at renewal, the whole team feels it. If the customer expands, the whole team celebrates. It eliminates the "not my job" mentality because the "job" is the relationship, not the task. This means you need to go beyond the typical key metrics by team, and also track project success rates.

There's always additional friction when you introduce this model and shared accountability. It will take trial and error to find the right incentive structure to align these teams toward the common goal of successful customer outcomes. When multiple people support an account, perspectives won't always align, and more communication effort will be needed early and often to ensure smooth execution.

The Feedback Loop: Closing the Gap

Once the right team is in place, with appropriate alignment, the next phase of defense for your Moat is the flow of information. In many startups, the CX team is a black hole. Customers scream, agents record tickets, CSMs fill out the product feature request form, and nothing changes. When humans are your Moat, CX is the business compass. Every ticket is a data point for the next version of your software.

You need a formal process to route feedback to your Product team. Depending on your stage, this could be a kanban board, a weekly sync, an Idea Portal, or a 2-day offsite. Pick the level of depth that is right for the stage your product is in, but do not limit yourself to bug or ticket review only. You need to listen to all customer feedback.

This loop only works if Product and CX share goals. If Product is measured on "Features Shipped" and CX is measured on "Tickets Closed," they will always be at war. You must align the incentives. When Product ships a feature that reduces support volume, Product should track and get credit for the retention win, too.

When all teams are aligned, retaining your customers becomes intuitive, because everyone is in the loop on what is working, where we need to focus, and what the priorities are. The better we become at keeping our customers, the more opportunity we naturally have to expand.

Enablement: The Knowledge Base as Force Multiplier

The Knowledge Base should not be a defensive tool to deflect tickets. It is an offensive tool to empower humans, including your internal teams and your external customer, and should always be created and managed by a human who leverages technology. A bad Knowledge Base is far, far worse than none at all.

Early in my management career, I made a mistake. I hired someone who was a bad fit for a support role. But I noticed something. The way she could document an issue was profound. She did not even need to fully understand a feature in order to describe it.

She became our first content strategist, and we quickly ended up with the best Knowledge Base in the business. By starting early, we were able to grow the Knowledge Base with the company, adding new products and features as we went. It set us up to scale both internally and externally every step of the way.

If you don't craft your Knowledge Base intelligently, it can be defensive, where the goal is to keep the customer out of the queue. When that is the goal, the content is often remedial. "Click here to reset password." This results in customer frustration and wasted time.

When designed as part of your Moat, the Knowledge Base takes on new life. It's not just customer education, it arms your people with context before they pick up the phone. Your knowledge should include internal articles and notes, and those should be updated at least as frequently as your external documentation.

Crucially, this is an area where you can and should use AI, but you should use it to empower your internal teams. AI can aid your content teams by translating technical release notes into customer friendly language drafts, which the content team can then review, update, and publish. If you load Knowledge Base content into systems that your AI agents can access, then your agents can become assistants *for the human agents*, not *for the customer.*

When deployed across your support ticketing system, you can also have your AI agents summarize and make suggestions while having your human agents read it, validate it, add necessary context, and apply empathy and humanity.

These are all examples of Digital Touch as a Force Multiplier. Look for ongoing areas to shorten cycle times, enable the humans to solve the problems, and allow you to scale intimacy without scaling headcount linearly.

The "Do More With Less" Mantra & Your AI Roadmap

Since the very beginnings of Customer Success, teams have been asked to do increasingly more with smaller and smaller teams. AI has exacerbated this trend, but AI has also made it so that this can finally be a real opportunity to do more with less while enhancing the customer experience, instead of detracting from it. When looking to do more with less, there are a couple of places it makes sense to start.

Do not look to AI to replace what's already successful. Instead, do an audit of your processes to determine what's currently breaking or missing across your team, tooling, and workflows. Where do your internal teams burn out? Where are you lacking visibility into your customers and their behavior? Where are the biggest hand off friction points?

If you are still unsure where to start, do a full customer journey mapping exercise. Note every stage that a customer goes through in their full lifecycle with you. Capture every handoff, what success looks like at each stage, and who owns successful navigation of that stage.

You'll find you do really well in some areas; leave those alone for now. You will find there are some areas where you do okay, but there is friction, or you know you could be better. You will also find there are areas where you are doing nothing. These are the areas where you should focus your initial efforts.

Once you understand your gaps, look at where AI may be able to overhaul or augment your existing team and processes in order to help. You should create an AI roadmap that allows you to clearly prioritize the items you intend to tackle, and in what order. Be sure to include costs, risks, and level of effort in your prioritization process.

Nearly all Customer Experience teams have significant gaps in Documentation, Support, or Data Management, so it is likely that you'll identify one or more of these as your biggest opportunity for improvement. Luckily, there is some low hanging fruit within each of these that AI is well positioned to solve.

Documentation: The Knowledge Base as a Living Asset

As we discussed, your knowledge base can be a huge differentiator, but in practice it's likely outdated. If your agents waste hours searching for answers and customers get frustrated by articles that don't match the current product, look for these as likely quick wins:

AI Knowledge Drafting: Use AI to draft, summarize, and update documentation. It can even suggest new articles based on support ticket types and common queries.

Release Notifications: AI can scan release notes and auto-draft simplified "What's New" emails for customers.

How-To Videos: AI can generate scripts for support agents to troubleshoot, scripts for CSMs to record demos, or summarize text articles into 30-second video snippets. Video AI programs can even create narrated videos and do video production on your behalf.

Search Optimization: AI can tag content so customers find the right article instantly, reducing ticket volume.

Data Management: Insights Over Information

You have data everywhere—structured (CRM, usage logs) and unstructured (support tickets, call transcripts) and it likely sits in silos. Your CSMs and Product Managers alike are drowning in spreadsheets but starving for insight. Use

AI here to synthesize and surface intelligence.
Trend Spotting: AI can scan thousands of tickets to identify everything from emerging bugs to recurring customer challenges. This lets you fix issues before they turn into churn.
Usage Trends: AI can correlate feature usage with churn risk to alert CSMs.
Action Items: AI can convert a call transcript into a structured follow-up task for the CSM ("Customer asked about Feature X, schedule a demo").
In-Call Insights: Real-time AI can listen to a call and prompt the CSM with relevant context ("This customer had a billing issue last month. Reference the credit you gave them.").

Support: The Triaging Layer

High volume, low complexity questions clog the queue. Your best agents are answering password reset tickets, and after-hours coverage is almost always both expensive and thin. Use AI for triage, not resolution.
After-Hours Support: AI can provide status updates and basic "how-to" answers when humans are asleep, ensuring the customer isn't left in silence.
Questions & How-To Support Response: A well configured AI Agent can handle 60% of repetitive queries (password, reset, navigation) instantly, and quickly reassign the other 40% to a human.
Data Gathering: AI can collect context (account ID, error code) before handing off to a human. This means when the human joins, the context for the problem is already 80% documented.
Phone, Email, Chat Data: AI can summarize the history so the human doesn't ask "What is your account number?"

Do not ever allow your AI to publish without validation, as AI can hallucinate technical details. A human expert should verify every AI-generated article before it goes live. AI can give the data, but the CSM must decide the intent. Do not automate conversations beyond the most basic of questions, use AI to prep the human, not to replace the voice. If the AI fails, the human must know everything the AI knows immediately. Never make the customer repeat themselves. The AI is a bridge, not a gate.

Chapter 6: Metrics That Matter

The Trap of the Vanity Metric

Many companies fall into the trap of measuring what is easy to measure, not what is meaningful. They track Ticket Volumes, First Response time, and satisfaction scores. These are often implemented with the best intentions, but can easily turn into vanity metrics whose purpose is to look good on a dashboard. Because they drive to a number instead of a behavior, they can easily be manipulated, and do not, by themselves, predict the health of your base.

Ticket Volumes, for example, can be lowered by making the product harder to use or the support harder to reach. That is not success even though it can look like it on paper.

First Response Times as well can be a trap. You can respond in 30 seconds with an automated message or a bot that doesn't solve the problem. That is not speed to resolution, it is simply speed as a measure all its own, without thought of benefit to the customer.

For NPS you can survey your happiest customers and ignore the quiet ones who are leaving. That is not showcasing customer loyalty, it's sampling bias.

I've seen all of those happen, and again, you have the opportunity to be different.

Your new Moat is not built on vanity. It is built on truth. You need metrics that measure intent, effort, and trust. These are the leading indicators that tell you if your Moat is holding or if the water is rising beneath the walls.

Let me be clear: this is not a call to stop tracking standard metrics. The typical metrics across CX are still very useful when given the appropriate context, and supported by the right teams and incentive structures. Rather, this is a call to recognize that if you want to be set apart, you need to go above and beyond what everybody else is doing. Tracking the basics just isn't enough.

Measuring Intent, Not Just Satisfaction

Unhappy customers don't always churn. Happy customers don't always renew. It's easy to rely on Customer Satisfaction Surveys (CSAT) and NPS as our measurements of customer success, but the true core of whether or not you will retain and expand your customer base is how many problems you solve for your customers, how many unmet needs they have, and how smooth you make it for them to go about their business.

Satisfaction is retrospective. Intent is prospective. If you ask a customer, "How satisfied are you?" they will tell you about the last interaction. If you ask, "What problems are we not currently solving for you?" or "Where do you see opportunities for us to provide the most value in next quarter?" you are measuring the business relationship. We always need to focus on what success means to the customer, because if the customer succeeds with your products, you succeed.

Still, how do you measure intent? Consider both Renewal Intent and Expansion Intent metrics. Customers who have positive intent indicators are more likely to expand and less likely to churn, regardless of their NPS.

Identifying Quiet Churn

The most dangerous signal is the absence of a signal. Most companies still wait for a renewal date to find out a customer is unhappy. By then, it is too late. You need a leading indicator of churn that does not rely on complaints.

We've all gotten that terrible cancellation notice from an enterprise customer. Out of the blue, because our data showed the customer was "Happy." NPS was 10/10. No recent support tickets. But login frequency had dropped 80% over 6 months. We did not notice because we were only tracking "success events." By the time the renewal came, the customer had quietly migrated to a new vendor. They didn't fight because they didn't think we cared. We didn't even notice until they were gone.

In the world of our Moat, silence is not golden. Silence is dangerous. A complaining customer is engaged. They care enough to fight. A silent customer has checked out. They are already mentally renewed with a competitor.

Defining Quiet Churn

Zero Activity: No login for 30 days.
Zero Contact: No support interaction for 90 days.
Adoption Decay: Feature usage dropping by 20% month-over-month.
The Trigger: If a customer meets 2 of these 3 criteria, they are "At Risk."

Use AI and other automation to alert your team when quiet churn triggers occur, but never automate the outreach itself. When your system flags the account due to these criteria being met, the CSM can reach out with a call and a personal note. Not a survey. Not a "how are you?" check-in. A human-to-human call. Followed by a specific value note.

Recover the silence before the renewal date. For most businesses, you should plan to execute your Save Plays between 3-6 months out from renewal, and have a solid understanding of who is unlikely to renew approximately a quarter before the contract expires.

The process can be similar for month to month contracts as well if they are large enough. Look at how long your typical customer stays. If it's 18 months, then at a minimum you want to start your At-Risk reviews at the 12 month mark, and have a solid understanding of your renewal status at the 15 month mark. If you have a high churn product, do a rolling assessment on a quarterly basis.

For low contract value customers who are low need you likely won't have a good sense of individual accounts. This is when you need to start looking hard at themes. Identify what successful, healthy customer behavior and usage patterns are, including both within your product and with the engagement they have with your company at events and online. Build At-Risk campaigns and communications that help drive your customers towards those behaviors.

The Cost of Effort

In the age of AI, if your customer has to work hard to work with you, learn your software, use your software, or get help with your software, you have failed.

Customer Effort Score (CES). This should be in every metrics deck, and hold more weight than NPS. High effort correlates directly with churn. Low effort correlates with advocacy. If you can reduce the effort your customers expend to achieve their goals, you make your product sticky. Empower your CSMs to provide their own CES assessment for their customers as well.

Go beyond just looking at the topline scores. The good always matters a whole lot less than the bad. Your goal is to measure the friction so you can fix it.

What percentage of tickets require more than one interaction to resolve? What percentage of AI interactions require a transfer from bot to human? (High here is bad; it means the AI isn't serving or enabling humans, it's just adding a step and additional friction).

Track both resolution time and rating on resolution, not just time to first response. The customer doesn't care when you pick up the phone. They care when the problem is fixed.

The Qualitative Moat: The Voice of the Customer

You cannot automate the relationship. In a data-driven world, we often forget the power of a human story. Your team should not just be closing tickets and checking the Quarterly Business Review (QBR) box, they should be gathering intelligence.

If people are your differentiator, you need to be deploying them at every level.

CSMs should have a "Talk" Metric, but so should leaders. Of course you should track QBRs, date of last touch, and other traditional metrics; however, you need to go beyond this as well. How many customers are your executives talking to weekly? Weekly, not monthly. How many feedback items are being logged? How quickly do we deliver on them?

How "heard" are our customers?

Much of this work is hand-to-hand, but do not underestimate the power of Conferences, Customer Advisory Boards (CABs) and Industry or Thought Leadership events.

View these not as feedback or complaint sessions, but as retention and expansion engines. Customers who sit on CABs have significantly lower churn and higher NRR because they feel ownership in the product. Align CAB membership with your highest performing and lowest performing segments to ensure you are listening to the right voices.

One of the most successful ways to start a CAB is by doing a feature specific group for the worst part of your product. Own the fact that it's below where you want it to be, and that you're eager to hear customer voices to make it wonderful. You'll find your angry customers become delighted to help solve the problem with you.

Conferences are often owned by Marketing, but Customer Success must own the Success Story track. Use conferences to showcase customer outcomes. This validates the value of your platform to prospects (Sales support) and current customers (Renewal confidence). This organically turns every conference appearance into a pipeline or retention opportunity.

Be sure to measure both the insight and the impact of the insight. Are you actually implementing what customers ask for? How often do you communicate that you've done so?

If you build a feature because a customer asked for it, tell them what you built and why you built it. This matters. It makes the customer feel that they matter. This feedback loop should be part of every deck, every CAB, every User Group.

The feedback loop often ends before it gets back to the customer. If you care about your customers, but you do not tell them you care, they will not feel it. You need to communicate early and often, and be transparent about your investment in your customers' success. Do not assume that because you announced a feature, your customers know about it, or know that it solves their problem. Build multiple communication channels into your processes to ensure customers are smothered with love and care.

AI & Metrics: The Radar Screen

As you integrate AI into your workflows, the way you measure success must evolve. AI can process data at a speed no human can match, but it cannot see the full picture. You must treat AI metrics as a radar screen, not the final word.

AI tools can now analyze thousands of support tickets to detect sentiment trends. They can tell you if frustration is rising across your entire base before it shows up in a survey. But AI can also misinterpret sarcasm or context. It might flag a happy customer as angry because of a specific keyword. Never act on sentiment data without human validation. Have a weekly "Sentiment Check" where a human reviews the AI's top 10 flagged interactions to confirm accuracy.

AI can predict churn probability based on usage patterns, but it can also give false positives. You might annoy a loyal customer by asking them if they are okay when they are fine and you should have known they were fine. Use the AI score as a priority list, not a trigger. If the AI says "High Churn Risk," the CSM investigates before reaching out. Do not send a generic "We noticed you're leaving" email. Ever.

AI metrics are efficient; human metrics are more likely to be accurate. Don't stop doing your quality checks after you launch or complete the initial training for a new AI. Conduct regular audits, more than you think you should.

The Value of Brand

I was chatting with a former customer turned friend, and she told me that while she loved some of the people at one of her current software providers, she didn't trust the company. It caused a real crisis for her emotionally, because she wanted to be honest and straightforward about some real concerns she had, but she didn't feel the company would take her feedback in good faith. She worried she'd have to deal with retaliatory hardball negotiation tactics just to keep her existing contract terms if she was honest. She won't be renewing.

Even if that company changed its policies today, it would take years for the changes to permeate throughout the organization, and for customers to really believe the business had changed. Meanwhile, customers are increasingly unhappy, starting to quietly leave, and are less and less likely to buy more products from the company.

That's the kind of brand reputation risk that really highlights the importance of humanity above policy. If your customers hate you, or even just mistrust you, everything becomes infinitely harder, especially at scale. That's part of the defense you can share in your board meetings. Brand reputation isn't fluff, it's a friendly referral, an open door to expansion, and a defense against competition.

The Board Deck Shift

In most startup board decks, Customer Experience is a footnote under "Operations." Worse yet, it's common for CX to not even be discussed in the Board Meeting if all the numbers are at or above plan. If your slides just rattle off Support Ticket trends, cost per ticket, NPS, etc, you're still speaking in the language of the cost center. It tells the Board you are efficient. It does not tell them you are valuable.

When the Moat is a human connection your board deck must reflect the value of that connection.

How do you report all of this without overwhelming leadership? It's not by including a bunch of new metrics. You likely have a mandated set of information from your board already. What you must do is ensure you have a narrative to share around leading, lagging, and intimacy numbers. When you've done the work internally in the teams to track the right data and signals, you have the information you need to craft this narrative at your fingertips.

It is your role to share the business successes and risks. If NRR is high but Effort is also high, you are on thin ice that won't show up if all you do is recite the retention numbers. You are relying on inertia. If Intimacy is low but NRR is high, you are at risk of a competitor disrupting you with a relationship.

What does this mean? It means you need to go a step further than your peers. Don't just report the "What." Give the "Why" and the "How." Your narrative insights are critical in showcasing the power of human connections to driving business value.

If you need a place to start, begin with highlighting what you've learned in recent Voice of Customer interviews, conversations, and cross-functional planning meetings. Talk about both positive and negative themes that you've seen in recent At Risk conversations, as well as what's been working and not working in your Save Plays. Customer stories have real power everywhere, including the board room.

These are unique value adds that contribute directly to your bottom line. Do not just present the numbers. Tell the story. If you are a Founder, you must lead this change. If you are a CS Leader, you must arm yourself with this data. You cannot rely on gut feeling. You must prove that you are both differentiated and that you drive business outcomes.

Use Your Board's Language

Investors and board members are not all created equal; however, they are always some of the most important relationships you will have. They will have a major impact on the type of company you build.

Understand the goals of your board, how they measure success, and how they are measured in their roles, and you will be better able to frame your human-centric work in a way that resonates.

If you have investors who are all about efficiency at all costs, you need to minimize discussions of Customer Experience costs, and highlight revenue growth from CX sources. Showcase expansion revenue generated from User Groups and Conferences. Highlight the reduced sales lifecycle time to cross-sell into existing customers as opposed to new customer acquisition. You'll find there are plenty of stories to tell to show the real value of being human first, but it's on you to craft a compelling narrative that will resonate with your audience.

Part III

Scaling the Moat

CHAPTER 7: DRIVING OUTCOMES

How Should I AI?

I was recently talking to a co-founder/CTO who I think has nailed the process for leveraging AI. He didn't start with a random productivity metric, he started with the problem he needed to solve. In his case it was QA, and the team required for the type of project he's running would typically be 20 people. Instead, he has a team of just two senior people who also run several AI agents to do the bulk of the work. The seniority of the people does two things. It means that they know what "good" looks like so they can set up the AI agents with very specific prompts, and it means they can continually reinforce the right behavior and guardrail against drift.

This structure allows his team to be lean, but it does something else, too. Because he has subject matter expert humans on his team overseeing this, he's protected if costs skyrocket and they need to pivot. Their AI isn't a black box, it's a tool they are using, led by expert hands. If they ever need to pivot they can easily do so, be it with additional people, different systems, or something else entirely.

The Build, Buy, or Borrow Decision

If you aren't using AI, you are falling behind. You need to stay up to date on the technology, and there's no better way to do it than to get your hands on it.

Once you identify where AI can help, you face the critical strategic choice of how to implement it. Ultimately, for each use case you wish to solve you'll need to choose to either build, buy, or borrow.

Regardless of which you select, make sure every AI initiative has an executive sponsor, a day to day owner, and clear metrics for success measurement.

Borrow: Leveraging AI already available in your existing stack.
Buy: Subscribing to a purpose-built point solution.
Build: Designing custom AI agents and workflows in-house.

Most companies default to "buy" because it's fast. Most founders default to "build" because they think it creates differentiation. The truth is, at the time of writing, 90% of AI should probably be borrowed or bought, with the majority borrowed.

Borrow: Using the AI features embedded in the tools you already pay for (CRM, CS Platform, Communication).
CRM: Salesforce Einstein, HubSpot AI, etc.
CS Platform: Gainsight Copilot, ChurnZero Insights, etc.
Communication: Zoom AI Assistant, Gong, Microsoft Teams & Copilot, etc.
Support: Zendesk AI, Agentforce, Intercom Fin, etc.

This functionality is ever changing, so spend time with your CSM at your software providers to understand up to date features and functionality.

Choose to borrow for maximum operational efficiency. It's a rare situation where you might be able to get significant improvements fast, cheap, and good, because these systems already have access to your customer and business data, and the incremental cost is likely low. In fact, check your contracts, you might already be paying for AI that you aren't using.

As with any additional software, you must still consider data privacy and limitations. Check where your data goes. Is it used to train public models and can you opt out? Bear in mind as well that these tools are designed for the *average* user and use case, not you specifically.

Have a back up plan that is human-led, so that you have options if prices go up or functionality disappears.

Buy: Subscribing to a standalone vendor for a specific capability that your borrowed tools lack.
Targeted Use Cases: Augmented Support, in-call coaching, and other agentic workflows are going to, at the time of writing, be much more robust when purchased as point solutions than when used as add-ons in existing systems.
Sentiment Analysis: Specialized tools that go deeper than standard NPS.
Advanced Triage: AI tools designed solely to route high-risk customers to specific humans.
Data Enrichment: Tools that pull external signals (funding news, hiring spikes) into your CSM's dashboard.

Choose this for critical capabilities or major gaps that have high impact if they go badly. Ensure what you buy integrates to your systems of record, and has ways to refresh and upload data.

Not all AI point solutions will stand the test of time. You must have a back up plan here as well that is human-led, so that you have options if prices go up or functionality disappears.

Build: Engineering custom AI agents or workflows that connect your unique data sources to specific customer outcomes.
Bespoke Needs: Custom designs can do exactly what you need them to do.
Proprietary Scoring: An AI model that combines product usage + billing history + support sentiment to predict *your* specific churn risk, which no vendor can offer.
Custom Workflows: An internal agent that triggers specific sequences based on a unique business logic you own, especially in niche but fairly stable industry and products.
Integration Hubs: Connecting your AI to your customer's specific data environment to offer advice only you can give.

Choose this if you have highly technical teams and enough proprietary info that your own internal AI insights can become a defensible advantage. If you build it, and it improves your NRR significantly, it becomes part of the Moat.

You also build yourself cost protects as well, particularly if you self host models and the applications you build on top of them.

Are you a company that has data scientists? Likely you'd be a good fit to do some build projects in-house. Do not think only about the build process, anything you develop will need maintenance, improvements, and fixes as time goes on. Will you have a dedicated team to build, test, and monitor?

Consider quick wins when getting started. If you can change the world in a year, but you can change a customer's least favorite interaction point in a week, do the interaction point first and get some positive proof points early. After some time, you'll likely have a combination of Build/Buy/Borrow, and will find some approaches work better than others for your team.

It's the Journey, not the Destination

In the early days of a startup, "success" is defined by survival. You keep the lights on and the coffee hot. You extinguish the immediate fire. Then, as you scale, survival is no longer enough. You've earned the privilege of needing to operationalize, and as part of that you must drive Predictability and Outcomes.

Many Customer Success teams define success as Renewal. If the customer renews, the team wins. This is a failure of ambition. Renewal is a lagging indicator. It is a receipt, not a strategy.

In a commoditized world, a customer who does not grow is a customer who waits for a competitor. They have no reason to stay except inertia.

The Moat is not just about retention. It is about naturally shortcutting to Value Realization. When you truly care about your customers, ensuring the customer achieves the business outcome they bought the software to achieve is job zero. When they achieve their outcomes, they stay, they advocate, and they expand.

The Usage Trap

To drive outcomes, you must first understand what "usage" actually means. In the SaaS industry there is a dangerous obsession with activity.

High usage does not equal high value. A customer can use your software every day to solve the wrong problem. A customer can spend hours configuring settings that add no value. They are using the tool, but they are not succeeding.

The other problem when you track usage as your primary indicator of health is that what naturally follows is measuring your team by activity.

This is common as you scale your team, especially when the founders had activities that worked well for them. But a newly hired team should never be tasked with the exact same metrics or activities *just because* they worked for the founding team. It's important to go a level deeper. Make sure you're building for the future, not just building on the past.

If your CSM can only report on activity, they are not driving outcomes. They are reporting on noise.

I had a customer who had 100% adoption of a critical feature. They were logging in, clicking buttons, and generating reports. Despite their textbook usage, their revenue was flat. They realized the tool was automating a process that didn't generate revenue. They stopped using the feature because it was costing them money. If the CS team had only tracked usage, they would have celebrated a "happy" customer. Instead, the customer churned.

If this is you, you need to shift your team's focus from *Activity* to *Outcome*.

The Outcome Playbook

How do you operationalize this? You need a Value Realization Plan (VRP) for every account. This is different from a Success Plan. A Success Plan is about your software and you should probably stop doing it. A Value Realization Plan is about helping the customer achieve their desired goals and outcomes for their business.

As always, the exact structure of this VRP will vary based on your business, but it needs to be customer goal oriented, have levers that relate to your product or service offerings, as well as include timelines, owners, and clear validation measurements.

This document is not a sales pitch. It is a contract of trust. It proves you care about their business, not just their renewal date.

Your QBRs should truly be treated as business reviews. Wherever possible, have an executive from your business participate, and whenever possible, do them in person. Try for at least annually.

Business Reviews should not be "How do you like our features?" They should not even be "Did we meet your business goals?" You should know whether you did or did not before you go into the call. The best Business Reviews are done in collaboration with the customer and are half proof-points of what you've accomplished, and half shared roadmap for what you are going to accomplish next. If the goal changes, the plan changes. And the goal will change. If it doesn't, that's a flag that you are out of the loop. Show that you are a partner rather than a vendor.

Scaling Intimacy: The Human Touch at Scale

If the economics worked out, I'd want to have a CSM for every 10-50 customers. In the real world; however, if you have 100,000 customers you may not be able to have a human CSM for every customer, even though you cannot afford to treat them like numbers either. There is a certain paradox that comes with scale.

Your challenge is to flip the script on customer segmentation. Don't blindly segment based on Average Contract Value (ACV) or territory or product. Look at customer and outcome complexity.

Your service levels may vary, but you'll likely want to structure with three general engagement models.

The highest level of service will be for your Strategic or Enterprise customer base. These are your high revenue (or potential for high revenue), high need, or high complexity customers. These customers need a dedicated human CSM, professional services and/or data contacts, and potentially priority support. These customers are invited to user conferences, CABs, and likely get regular product roadmap updates and access to senior leaders.

The next level of service is for your Growth customers. These accounts might be low revenue, with medium or high complexity, medium revenue with medium complexity, or another similar combination. These customers also need a dedicated CSM, but likely need less frequent and less complex communications, and standard support is likely sufficient. These customers are also invited to user conferences, may be invited to CABs, and get big picture updates on strategic direction periodically.

The lowest level of service is for your lowest revenue customers who also have low need and low complexity. These customers may not have a dedicated CSM, but should still receive proactive engagement, along with standard support. These customers are invited to user conferences as well, may be invited to CABs, and get big picture updates on strategic direction periodically.

In this model, there is no "digital only" level of service. Even our lowest-tier customers require a human behind the wheel running campaigns. The Moat requires human presence at every level, even if that presence is scaled differently.

Notice that all customers should be invited to user conferences, CABs, product updates, and other events. Events can be an absolute superpower for energizing your customer base, and you shouldn't keep that goodness away from your smaller customers. You may need to have separate content tracks, different topics in your CABs, and other targeted approaches, but open your event doors wide enough to let all of the customers you serve get value.

Hybrid Execution

For all customers, you should use technology to enable the human touch. That's right, your Digital and Scaled program should not be limited to only your small customers. Build your program in a way that adds value to your entire customer base.

AI or other automation should analyze the customer's data and guide the CSM toward successful and struggling customers with some personalized insights. This lets CSMs better prioritize their time, the customer feels seen, the CSM saves time, the relationship scales. When CSMs plan their efforts based on data, they know *who* to call and *when*.

Note that I said "call" again. In this increasingly digital world, picking up the phone is more powerful than ever.

Digital is not "automation replacing humans." It is AI surfacing the signal so humans can amplify it.

The Expansion Engine

When you focus on doing the right thing, caring for your customer, and delivering on outcomes, expansion becomes a byproduct.
In traditional sales, expansion is "selling more seats." Expansion is the natural progression of doing right by the customer. This shift is critical. It moves the conversation from Price to Solution.

Scaling Without Breaking Culture

A secret challenge when building a successful company is that what got you here won't get you to the next level. You might have built beautiful processes that delight your first 10 customers. You cannot use that process for 1,000.
As you grow, ensure you are structured in a way that allows for revision and replication.

The "Who We Are" Check

Change is part of startup life, so don't shy away from it. Focus on the most important things first, and then move on to the next. Every time you scale a process, ask: "Does this feel like who we are?"

Chapter 8: The Future of CX

The Horizon is Human

Technology moves faster than strategy. As we look to the next quarter, next year, next decade, the question is no longer "Will AI change Customer Experience?" The question is "How do we build a Moat that survives the change?"

Many leaders fear the future. They see AI as a threat to their business model, or conversely, as a magic wand to cut costs. Both views are dangerous. If you see AI as a threat, you will hide from it. If you see it as a magic wand, you will burn your foundation to ash.

The future of CX is not Human vs. AI. It is Human *with* technology, just as it has always been. But humanity has become a nice-to-have for too many businesses, and this is your opportunity to capitalize on the use of technology to serve you and your customers.

AI is the Radar: It scans the horizon, processes the data, and detects the storm. Your people are the crew. They interpret the signal, decide the course, and steer the ship.

Future proofing your team is about equipping your humans better than ever before, armed with the best technology available, but grounded in the one thing code cannot replicate: empathy.

AI as the Augmentation Engine

AI and Technology are excellent enablers as co-pilots to your team. Imagine a CSM managing 150 accounts. Without AI, they can only check in when they remember. With AI, they have a dashboard that lights up. This can radically change the way a CSM does their work, by systematically targeting the most important and urgent work.

There is an inherent danger of relying too heavily on the machine. If you don't understand how the AI reached a conclusion, you cannot explain it to a customer.

Don't abdicate your thinking to technology. Let it sort through the noise, point you in the right direction, and do the rote tasks for you, but use your human discernment to validate and go the extra step. Do this more frequently than you think you need to, and build thinking time into your leaders and your front line team members processes. It's a perishable skill.

The New Skill Set: Hiring for the Future

If the AI handles most of the data, what do you hire humans to do? The skill set required for human-centered teams is shifting.

From Technical to Strategic

In the past, a support agent needed to know the software inside and out. In the future, they need to know the customer inside and out.

AI can memorize your product knowledge base. Diagnosing the business problem behind the technical request? That's still solidly in the realm of human skills.

The best human traits are still the best human traits. Look for people who can navigate the grey areas with empathy, approach their work with curiosity, learn fast, are resilient, adaptable, and have a bias for action.

The Transparency Imperative

As AI becomes more prevalent in your customer interactions, Transparency becomes your new currency. Customers are smart. They can tell when they are talking to a bot.

The "Human Reviewed" Standard

Do not try to pass off AI interactions as human. It is a short-term gain for a long-term loss. If a bot is involved, say so. "I'm using AI tools to find this answer faster." Ensure there is always a clear path to a human. "If this doesn't resolve it, I will escalate this personally."

Paradoxically, admitting you use AI builds trust. It shows you are efficient but honest.

Data privacy and security cannot be left out of the equation. When you use AI, you must ensure customer data is not being used to train public models without consent.

Future-Proofing the Team

Scaling your Moat means your team must evolve as fast as the technology. This requires a commitment to continuous learning that is woven into the fabric of the work, not just a quarterly seminar.

Think of continuous learning as the new easy button for continuous improvement. This has the potential to be the heartbeat of your organization. Consider non-technical team hackathons with AI and other emerging technology. This style of experimentation will not replace their thinking, but sharpen it. They need to know how to prompt, how to verify, and how to use the new capabilities safely. Be aware that this cannot happen in a vacuum.

There must also be a consistent focus on customer interactions and learnings. This means listening to calls, reading tickets, and feeling pain points directly. This keeps the team grounded in reality.

Strategic offsites have never been more important. If possible, bring teams together quarterly, but at a minimum carve out meaningful virtual time quarterly, and annually in person. Strategy sessions let you pull up from the day to day and allow you to focus on business outcomes rather than tools or today's fire. It is where you remind everyone why we are doing this in the first place.

The Friction Point: Resistance to Change

There will be resistance. You cannot assume your team will welcome the constant change of new tools with open arms, even when it's an objective improvement. Some will not be able to overcome the change fatigue, some will fear AI, and some will try to use it to do less work. Don't underestimate the human element of the transformation.

Reward open communication, alignment and adoption. Remove barriers. If a team member finds a way to use systems better, celebrate it publicly. Why?

Because there is real risk in not just going too fast, but in going too slowly. Have a clear perspective on approved and not approved tooling and usage so that you can minimize shadow IT risk. If your team starts using unauthorized AI tools to get work done because the sanctioned tools are too slow, you risk data leaks.

You must provide sanctioned tools and give them to the people who need them. If they need more, give them more. The culture must trust your team to use the tools right. Do not punish experimentation. If someone tries a new workflow and it works, adopt it. If it fails, learn from it. Trust your team to use the tools right.

The AI Enabled Team

You need to ensure you are not just buying technology, you are building capability. It proves that your Moat is not static, but that it grows stronger as the technology evolves.

How do you measure your readiness, capability, and progress? You cannot just rely on the old metrics of volume and speed. Stop looking only for numbers, and begin looking for the story of your capability.

You want to see a high Adoption Rate of your new technology, meaning a percentage of tools are being used effectively by the team. You also want to know how often humans override AI suggestions. If this number is too low, it means your team has stopped thinking. It means they are just rubber-stamping the machine.

As your systems and tooling start enabling your teams to do more, track Skill Growth, measuring how many team members are certified in new technology including AI, systems, and advanced features. And of course, track what impact these new competencies have on your business outcomes.

The future is bright. But only if you steer it with a human hand.

CHAPTER 9: KEEPING THE SOUL

The Long View

This book began in the shadow of the new. It started with the fear that Artificial Intelligence was coming to erase our advantage; startups rendered obsolete overnight. It felt like the game was rigged against us.

Now, there is a Moat and a map of the future. Now at the conclusion, I want to ask you one simple question:

When you close this book, what is the one thing you will remember?

I hope it is not the metric definitions or the different ways you can structure a team. I hope it is the truth uncovered at the very beginning: in a world of infinite intelligence, human connection is the only distinguishing feature that actually matters.

The "Early Days" Core

I often think back to the beginnings at the earliest stage startup I ever joined. Before we were PE-backed. Before we were acquired. We were small. We were unpolished. Our product had bugs. Our uptime was not the highest. But we were hungry. We remembered what it felt like to be answering support emails at 10:00 PM on Thanksgiving because a customer couldn't get on the platform.

Those memories are just a few of hundreds because that mindset was the soul of the business.

Years later, when the acquisition came, our new owners looked at our code. They saw gaps. They saw technical debt. But they also looked at our customer list. They saw retention that defied industry logic. They asked me, "How did you keep them?"

I told them the truth: "We have a personality, and we're not afraid to show we care. We're people they want to work with, not just a software vendor."

We had built a Moat based on human connection. And customers didn't just buy our product. They bought our personal approach. They bought the relationships we had cultivated over time. That is the only asset that appreciates with age. Code depreciates. Relationships compound.

The Core Thesis, Revisited

Let us be clear about what we are building as strategic software leaders. We are not building a support department. We are not building a customer success team. We are building a defensible Moat.

The software industry has convinced us that the goal is efficiency. It is the language of the spreadsheet. It is the logic of the boardroom. Minimize cost. Maximize speed. Deflect tickets. Unfortunately, this logic is flawed. It assumes that the customer is a number to be processed.

I want you to challenge that logic.

Efficiency is not the goal. **Effectiveness is.** Speed is not the goal. **Resolution is.** Cost is not the goal. **Value is.**

When you prioritize efficiency over effectiveness, you build a faster path to churn. When you prioritize cost over value, you build a cheaper path to irrelevance. To truly be your differentiator, the Moat requires you to make the hard choice. It requires you to invest in people when the spreadsheet says you shouldn't. It requires you to answer the phone when the bot says "no."

The Personal Cost of Convenience

There is a toll paid by the founders who strip the humans out of their business to show margin to investors. There is a toll paid by the employees who are forced to hide behind scripts because their company fears accountability.

I have seen teams burn out because they were expected to be machines. I have seen founders wake up one day with a million in revenue but a customer base that hates them. They have the money, but they have lost the market. They have the asset, but they have lost the heart.

This is why keeping the soul is not just a moral decision. It is a business imperative.

As a leader, your stress is lower when you don't have to apologize for the company's actions. When your team is empowered, you sleep better. Your employee engagement is higher when you empower them to solve problems. They are not paid to read scripts. They are paid to use your brain and your heart.

And the customer's life is better. No more shouting into a void. They can be secure in the knowledge that they are partnering with someone who cares.

The Differentiator of Tomorrow

As we look at the next ten years, the competition will not be other software. The competition will be everything else. Your customer will be able to solve their problem with AI agents. They will be able to write their own workflows. They will be able to clone your features.

So what is left? The relationship.

When the tool fails—and it will fail—your customers will call you. If you are automated, they will be met with silence. They will leave. If you are human and you meet them with care, they will stay.

When you exit (via VC, PE, IPO or M&A), the buyer isn't just buying ARR. They are buying the stability of your retention and the loyalty of your base. A company without a soul has nothing it can defend.

That is the differentiator. It is not a feature. It is a feeling. It is the knowledge that when the storm comes, there is a human on the other end of the line who will help you navigate it.

This is the only Moat that cannot be replicated. AI can learn to mimic empathy, but it cannot feel it. AI can simulate care, but it cannot choose it.

The Leader's Responsibility

This falls on you. If you are reading this, you are likely a founder, an executive, or a customer experience leader. You hold the pen that writes the check for the team. You hold the voice that tells the Board what matters.

It is hard. It is messy. It is expensive. It is the only way to win the long game.

Do not outsource your responsibility to the algorithm. Protect your culture like your life depends on it. Because in a sense, your business life does.

Part IV

Appendix

The original intent of this book was to provide only the theory, and leave the nuance of implementing the concepts to the reader. Knowledge without execution; however, is a liability in an era of AI commoditization. It's easy to read the ideals in the previous pages, enjoy the frameworks, and then slip smoothly into old habits.

I have included these worksheets as an execution forcing function.

Context is key when considering your responses. If your model is high-volume/low-touch, certain frameworks may feel heavy. If you operate in high-touch/enterprise territory, others may feel light. Do not treat these worksheets as gospel.

Instead, use these tools to carve out time for the hard work of self-assessment and the even harder work of change. All of these can be done solo or with your leadership team as an offsite strategic exercise.

Do not outsource your strategic thinking to me or these pages, use them to synthesize your perspective and challenge your assumptions.

Worksheet 1: Maturity Assessment

Instructions: Answer the assessment honestly by selecting the option that reflects the current state across the core areas of your business.

STRATEGY & VISION

1. Our Board treats CX as a strategic asset.

[ ] 1 pt: CX discussed only when tickets/churn spike.

[ ] 3 pts: CX discussed quarterly as a metric, but not primary KPIs.

[ ] 5 pts: CX reports on KPIs in all board decks, tied to valuation.

2. Hiring process screens for empathy/judgment.

[ ] 1 pt: Hiring is based on technical fit and experience level.

[ ] 3 pts: Hiring includes behavioral/culture fit questions, but tech dominates.

[ ] 5 pts: Successful candidates must pass both technical and behavior/culture fit assessments.

3. Founders/Execs spend time on the front lines regularly.

[ ] 1 pt: Leadership is not involved in support or customer operations.

[ ] 3 pts: Leadership listens to calls once a quarter or when customers escalate.

[ ] 5 pts: Leadership has a recurring, scheduled cadence or live, in person, and recorded customer interactions.

4. Budget allocates for "delight," not just efficiency.

[ ] 1 pt: Support budget is cut when margins need improvement.

[ ] 3 pts: Budget is fixed annually by management only.

[ ] 5 pts: Budget includes a "relationship investment" line item with flexibility for individual contributors to leverage.

5. Growth relies on retention/expansion, not just acquisition.

[ ] 1 pt: Focus is primarily on new logo acquisition.

[ ] 3 pts: Retention is tracked, but acquisition is the main lever.

[ ] 5 pts: Retention/NRR is the primary engine; acquisition supports it.

OPERATIONS & STRUCTURE

6. Organizational structure minimizes handoffs.

[ ] 1 pt: Strict silos (Sales ⊠ Success ⊠ Support).

[ ] 3 pts: Some cross-functional pods, broken handoffs are common.

[ ] 5 pts: Clear, unified point/s of contact from sale to renewal.

7. Employees have authority to solve problems without approval.

[ ] 1 pt: Every decision requires escalation.

[ ] 3 pts: Managers are available for quick approvals.

[ ] 5 pts: Agents have defined discretion within preset limits.

8. Knowledge Base focuses on satisfied resolution, not deflections.

[ ] 1 pt: KB is designed for customers to read before calling.

[ ] 3 pts: KB has internal notes for agents, in addition to external content.

[ ] 5 pts: AI-enabled, up-to-date KB drafts solutions for internal and external engagement.

9. Formal feedback loop where CX influences Product roadmaps.

[ ] 1 pt: Product builds what execs/investors think is cool.

[ ] 3 pts: Product reviews tickets monthly w/limited support focused resources.

[ ] 5 pts: Product/CX has regular cadences for feedback at multiple levels throughout the organization.

10. Onboarding exposes new employees to customers.

[ ] 1 pt: Onboarding is payroll, benefits, HR policies.

[ ] 3 pts: Onboarding includes a product demo.

[ ] 5 pts: Onboarding includes Customer Contract and listening to calls/working with a customer.

METRICS & VALUE

11. In Support, Retention is tracked as a primary KPI.

[ ] 1 pt: Response time is the primary dashboard metric.

[ ] 3 pts: Response time and Retention are tracked with no clear hierarchy.

[ ] 5 pts: Retention and Satisfaction are primary; Response time secondary.

12. Customer Effort Score (CES) or similar "Outcome" metrics are tracked.

[ ] 1 pt: We only measure CSAT.

[ ] 3 pts: We measure CES occasionally but rarely take action on it.

[ ] 5 pts: CES drives strategic goals and team incentives.

13. Identify "Quiet Churn" risks before customers leave.

[ ] 1 pt: We wait for a renewal meeting to find unhappiness.

[ ] 3 pts: We flag low activity, but rarely intervene.

[ ] 5 pts: We have a proactive system in place for silent accounts.

14. "Intent Signals" are tracked.

[ ] 1 pt: We track usage/volume only.

[ ] 3 pts: We track usage and do CSM check-ins.

[ ] 5 pts: We score customers on "Intent Signals" as part of health and other KPI metrics.

15. The Board deck tells a story of value, not just efficiency.

[ ] 1 pt: We focus on ticket deflection and cost.

[ ] 3 pts: We include NPS and Churn in Board decks, but don't review if the numbers are good.

[ ] 5 pts: We include "Value Realization" stories & Risk Themes along with other CX metrics in board meetings.

CX & CULTURE

16. The team is rewarded for outcomes, not volume.

[ ] 1 pt: Comp is tied strictly to speed/volume.

[ ] 3 pts: Comp includes retention incentives for managers only.

[ ] 5 pts: Comp for all CSMs tied to Retention and Outcomes.

17. Customers can reach a human easily when the system fails.

[ ] 1 pt: Support is accessible behind automation layers.

[ ] 3 pts: Support is accessible for critical issues, but wait times vary.

[ ] 5 pts: We provide guaranteed paths to humans within minutes.

18. Celebrate "wins" where we helped a customer achieve goals.

[ ] 1 pt: We celebrate closing sales.

[ ] 3 pts: We celebrate closing sales and renewals.

[ ] 5 pts: We celebrate customer outcomes.

19. We celebrate "inefficiency" if it solves critical customer problems.

[ ] 1 pt: We prioritize speed and process compliance.

[ ] 3 pts: With manager approval, we make exceptions for VIP accounts.

[ ] 5 pts: We empower the team to do right by our customers, even if that memes bending process.

20. Culture prevents burnout among support staff.

[ ] 1 pt: High attrition, no promotions, low engagement.

[ ] 3 pts: Average attrition, some promotions, average engagement.

[ ] 5 pts: Low attrition, high promotion, high engagement.

SCORING GUIDE

Total Score: ______ / 100

20-49: You are likely vulnerable to AI commoditization. Look at your lowest scoring areas and identify 2 or 3 quick wins to immediately address.

50-74: Your foundation exists, but behaviors are inconsistent. Look at your lowest scoring areas and build a roadmap to prioritize addressing all areas at a 1.

75+: You have built the Moat! Maintain focus on relationships and continue investing in your team and your customers.

Worksheet 2: The Customer Contract

Instructions: Define your operating principles by answering these questions honestly to determine who you are, and reflect on who you wish to be.

DIAGNOSE THE FRICTION

Q1: When a critical issue arises, what happens first?

Q2: When a customer has a request that is outside policy, what is the default answer?

Q3: What is the most common complaint we receive?

Q4: If a manager has to choose between metrics or a happy customer, what wins?

DEFINE YOUR CORE PRINCIPLES

Draft 3–5 Core Principles that guide the core values you want your team to embody.

Example: Responsibility

When things break, who is accountable?

Example: Communication

How do we deliver good news? Bad news? How fast?

Principle 1:

Principle 2:

Principle 3:

Principle 4:

Principle 5:

IMPLEMENTATION

Principles are useless without some structure and rules. Write one specific "Always/Never" rule for each.

Principle 1:

Our "Always":

Our "Never":

Principle 2:

Our "Always":

Our "Never":

Principle 3:

Our "Always":

Our "Never":

Principle 4:

Our "Always":

Our "Never":

Principle 5:

Our "Always":

Our "Never":

NOTES FOR IMPLEMENTATION

Draft: Take your new principles and draft them into the format that makes sense for your culture. For some, this is short themes and statements. For others, this is more narrative in form. The right way to do it is the way that will resonate best with your team.

Display: Once complete. post this on the wall in your building entrance, your conference rooms, and shared systems.

Review: Add this to the agenda in customer calls. "Did we live up to this contract this quarter?"

Worksheet 3: 360° Dashboard

Instructions: Copy this structure into your BI tool or system of record and review along with all other core business metrics.

SECTION 1: HEALTH (Foundational)

Metric	Target	Current	Status
NRR	> 115%		
GRR	> 95%		
Logo Churn	< 2%/mo		
Rev Churn	< 5%/yr		
CAC Payback	< 12 mos		

If NRR is high but so is Churn, identify and fix the leaks. If Churn is low but NRR is less than 100%, focus on Outcomes blockers to Customer Expansion

SECTION 2: EFFORT (Leading Indicators)

Metric	Target	Current	Status
CES	> 4.5		
Time to Resolution	< 24 hrs		
Escalation Rate	< 10%		
Silent Churn Risk	< 5%		

If CES is dropping but your NRR is flat or also dropping, you are losing loyalty. Dig in and investigate.

SECTION 3: INTIMACY (Relationship)

Metric	Target	Current	Status
Intent Signals	> 60%		
Feedback Response Rate	> 20%		
Advocacy Count (Referrals, Speaking, Etc.)	> 10/mo		
Senior Customer Contacts	>50% per CSM		
Empowerment Budget Usage	Trending Up		

If Advocacy is low, customers are happy but not loyal. Pay attention to both the raw numbers and the trends.

SECTION 4: SUCCESS STORIES

Attempting to fill this out without the help of any of your individual contributors will give you great insight into just how close you are to your customers as a leadership team.

One Win: Example: "We helped Customer A achieve Outcome B this quarter."
Our Win:

One Risk: Example: "We identified Risk C in the Silent Churn metric, executed on Feature D and Retained $X ARR."

Our Risk:

One Insight: Example: "Customers were asking for Feature A to solve Problem B. We identified Function already in Product as a solution and new use case. This will unlock $X Retained ARR, $X Expansion ARR and $X New Business ARR."

Our Insight:

Worksheet 4: Value Realization Plan

Instructions: Create during onboarding and reassess twice yearly. Build in concert with the customer.

SECTION 1: DISCOVERY (The "Why")

Discovery Question	Customer Response
1. What's the one thing keeping you up at night?	
2. What does success look like in 12 months?	
3. What are the main risks if this project fails?	
4. Who benefits internally if we succeed?	
5. If this were to fail, what would be the most likely reason?	
6. What are the critical success factors? (Customer and us)	
7. What are the resource requirements? (Customer and us)	

SECTION 2: THE PLAN (The "How")

Goal / Target Date	Lever	Actions / Owners
Ex: Reduce costs 30% by Q2	New K- Base	Train agents

SECTION 3: RISK ASSESSMENT

Potential Risk	Probability (L/M/H)	Impact (L/M/H)	Mitigation
Ex: Champion leaves			

SECTION 4: SHARED AGENDA (30 Mins)

Depending on your product and service, this might be a portion of your QBR agenda, might be a separate meeting, or it may replace the QBR.

Review Mission: Are we still aligned on the goal? (5 mins)

Check Metrics: Did we hit the agreed upon metrics? What should we celebrate? Where is there still work to do? (10 mins)

Review Risks: Are there any current risks? For known risks, are the mitigation strategies working? (5 mins)

Plan Next Steps: Align on priority for next quarter (10 mins)

WORKSHEET 5: ORG DESIGN

Instructions: Assess your business using the questions below. Of all the worksheets in this book, this will have the most variance depending on your organization, so use this as a thought exercise, don't approach assuming you must end up in one specific model.

STEP 1: ASSESS YOUR CUSTOMER PROFILE

Q1. Average Contract Value (ACV)?

[ ] A. Low Revenue

[ ] B. Mid Market

[ ] C. Enterprise/Strategic

Q2. Implementation Complexity?

[ ] A. Low (<1 day, Self Service, Plug & Play)

[ ] B. Medium (1–4 weeks, Training required)

[ ] C. High (>1 month, Professional Services required)

Q3. Strategic Guidance Needed?

[ ] A. Low (Customer has the knowledge, they just need the product)

[ ] B. Medium (Integration or workflow guidance needed)

[ ] C. High (Process improvement, change management, data transformation, and/or high complexity)

STEP 2: DETERMINE YOUR MODEL

If you chose mostly A:

Model: Digital First / Community

Structure: Community, digital, and group (webinars, how-to emails, office hours, etc.) management approach led by a named person or team.

Human Touch: "Human on Demand." Customers have a path to contact a person when needed.

Focus: Efficiency at scale.

If you chose mostly B:

Model: CSM

Structure: One CSM owns account Sale ⊠ Renewal.

Human Touch: Weekly/Monthly video/email check-ins.

Focus: Relationship + Adoption.

If you chose mostly C:

Model: The Team "Pod"

Structure: Team owns account (PM + CSM + Support + Product Liaison).

Human Touch: Dedicated Slack, QBRs, Executive Sponsorship.

Focus: Strategic Outcomes.

If you have a Mix:

Model: Hybrid

Structure: Align teams to Digital/CSM/Pod based on segment.

Human Touch: Will vary based on the model assigned to the segment.

Focus: Targeted by segment.

Worksheet 6: AI Audit

Instructions: Answer the assessment honestly by capturing your current AI footprint, your behavior around it, and identifying risk and appropriate mitigations.

SECTION 1: TRANSPARENCY & TRUST

QUESTION	Y/N	RISK IF NO	PLAN
1. Disclose when AI agent is involved?			
2. Clear, one-click path to a human?			
3. Train customers on AI tools?			
4. Employees feel their jobs are safe?			

SECTION 2: DATA PRIVACY & ETHICS

QUESTION	Y/N	RISK IF NO	PLAN
5. Policy on data use for model training?			
6. Opt-out for AI analysis of data?			
7. Auditing AI for bias?			
8. AI logs stored securely?			

SECTION 3: AUGMENTATION QUALITY

QUESTION	Y/N	RISK IF NO	PLAN
9. AI summaries for internal team?			
10. Internal validation of AI responses?			
11. Measurement of human validation/ rate?			
12. AI helps surface relevant knowledge info?			

SECTION 4: HUMAN SAFETY NET

QUESTION	Y/N	RISK IF NO	PLAN
13. Protocol in place for high-risk customers?			
14. Training on how to review AI insights?			
15. "Kill switch" for unpredictable AI?			

HUMAN VALIDATION RULE:

Do not underinvest in your validation process. At a minimum, plan to conduct 1 human spot check for every 10 AI insights.

Assign a Senior Customer Success team member to also audit 10 random AI responses per week.

Worksheet 7: Customer Connection

Instructions: Set a recurring meeting for yourself as a leader to stay close to your customers by directly engaging with them. Select at least 1 of the following actions per week.

Listen

Listen to three random support calls.

Do not flag them. Just listen.

Learn

Ask your CSMs: "What is the one thing the customer is asking for that we are saying no to?"

Verify

Look at your NPS comments and deep dive into the feedback behind recent Passives and Detractors.

Connect

Reach out and speak to at least 1 customer. Avoid your biggest, happiest, or friendliest customer. Tell them you want to understand their challenges. Then, just listen.

About the Author

Jennifer Courchaine is the author of *The Moat*, a definitive guide to building human-centric businesses in the age of AI. With nearly 20 years of Customer Experience leadership, Jennifer specializes in scaling technology companies through angel investment, venture funding, and strategic exits across Legal Tech, Risk Management, Automotive, and Nonprofit sectors.

A thought leader in the CX space, Jennifer holds a Master of Arts in Strategic Communication and has served as Chair of the industry-prestigious PREX and XChange conferences. She is a frequent conference speaker and startup advisor who advocates for a product mindset that prioritizes relationships over efficiency at all cost.

Based in Vancouver, Washington, Jennifer lives by the same principles she teaches: build strong teams, invest in growth, and stay connected to your community. When she isn't scaling organizations or advising founders, she is exploring the Pacific Northwest as a Private Pilot, teaching SCUBA as a PADI Open Water Instructor, or perfecting a new dessert recipe. She travels frequently with her husband, Charles, and is currently planning her visit to the next continent on her list.

www.ingramcontent.com/pod-product-compliance
Lightning Source LLC
La Vergne TN
LVHW010840120826
845149LV00017B/3337

9798995360407